SAE EDGE™
RESEARCH REPORT

Unsettled Technology Domains in Aerospace Additive Manufacturing Concerning Safety, Airworthiness, and Certification

Dr. Jody E. Muelaner
Muelaner Engineering Ltd.

EDGE DEVELOPMENT TEAM

Dr. Nadia Vialas,
Head of Materials and Processes, Liebherr-Aerospace

Alexander Altmann,
Head of AM R&T, Liebherr-Aerospace

Claudia Jordan,
EOS GmbH Electro Optical Systems

Chet Fuller,
Chief Commercial Officer of Norsk Titanium

Robin Weston,
AM Marketing & Product Manager, Renishaw

Jonathan Porter,
AM Business Development Manager, Renishaw

Dr. Bhaskar Dutta, *DM3D Technology*

Dr. Joseph Flynn,
Assistant Professor (Lecturer), University of Bath

John O. Milewski,
Los Alamos Nat. Lab. (retired), APEX3D LLC

SAE INTERNATIONAL®

Warrendale, Pennsylvania, USA

About the Publisher

SAE International® is a global association of more than 128,000 engineers and related technical experts in the aerospace, automotive, and commercial-vehicle industries. Our core competencies are lifelong learning and voluntary consensus standards development. Visit sae.org

SAE EDGE™ Research Report Disclaimer

SAE EDGE™ Research Reports focus on topics that are dynamic, in which knowledge is incomplete, and which have yet to be standardized. They represent the collective wisdom of a group of experts and serve as a practical guide to the reader in understanding unsettled subject matter. They are not meant to provide a recommended practice or protocol. The experts engaged have contributed their own thoughts and points of view, and these are not the positions of the institutions or businesses with which they are affiliated. A professional writer has collectivized their input; there is no one contributor's perspective being advanced, but rather that of a community of practitioners. SAE EDGE™ Research Reports are the property of SAE International, and SAE alone is responsible for their content.

About This Publication

SAE EDGE™ Research Reports provide state-of-the-art and state-of-the-industry examinations of the most significant topics in mobility engineering. SAE EDGE™ contributors are experts from research, academia, and industry who have come together to explore and define the most critical advancements, challenges, and future direction in areas such as vehicle automation, unmanned aircraft, Internet of Things and connectivity, cybersecurity, advanced propulsion, and advanced manufacturing.

Related Resources

SAE MOBILUS® Advanced Manufacturing Knowledge Hub
https://saemobilus.sae.org/advanced-manufacturing/

SAE Team

Frank Menchaca, Chief Product Officer
Michael Thompson, Director, Standards, Information and Research Publications
Monica Nogueira, Acquisitions Director
Jill Leonard, Product Manager
William Kucinski, Managing Technical Editor

Printed in USA

ISSN 2640-3536
e-ISSN 2640-3544
ISBN 978-1-4686-0119-0

To purchase bulk quantities, please contact: SAE Customer Service

E-mail: CustomerService@sae.org
Phone: 877-606-7323 (*inside USA and Canada*)
724-776-4970 (*outside USA*)
Fax: 724-776-0790

Visit the SAE International Bookstore at books.sae.org

About the Author

Dr. Jody E. Muelaner is a chartered mechanical engineer with a background in metrology, aerospace manufacturing, and machine design. He now specializes in writing about technical topics in a way that the target audience can easily understand.

His writing has included technical reports for Rolls-Royce and Airbus, peer-reviewed journals, and UK Government reports, as well as magazines and websites. He has published several hundred articles and was awarded the Sage Best Paper Award in 2010.

Starting out in machine design, Dr. Muelaner initially worked on sawmills, waste-processing machinery, domestic appliances, and medical devices. After moving into metrology, his research focused on modelling and optimizing uncertainty in manufacturing systems, enabling right-first-time assembly, and the design of innovative laser instruments. He founded Muelaner Engineering Ltd in 2018 to provide consultancy and technical writing within advanced manufacturing and machine design.

contents

Atosan/Shutterstock.com

Unsettled Technology Domains in Aerospace Additive Manufacturing Concerning Safety, Airworthiness, and Certification

DR. JODY E. MUELANER

Muelaner Engineering Ltd.

Abstract

Additive manufacturing (AM) is currently being used to produce many certified aerospace components. However, significant advantages of AM are not exploited due to unresolved issues associated with process control, feedstock materials, surface finish, inspection, and cost. Components subject to fatigue must undergo surface finish improvements to enable inspection. This adds cost and limits the use of topology optimization. Continued development of process models is also required to enable optimization and understand the potential for defects in thin-walled and slender sections. Costs are high for powder-fed processes due to material costs, machine costs, and low deposition rates. Costs for wire-fed processes are high due to the extensive postprocess machining required. In addition, these processes are limited to low-complexity features. Incremental improvements in all of these areas are being made, but a step change could potentially be achieved by hybrid processes, which use wire feedstock to deposit the bulk of the part and powder for fine detail.

NOTE: SAE EDGE™ Research Reports are intended to identify and illuminate key issues in emerging, but still unsettled, technologies of interest to the mobility industry. The goal of SAE EDGE™ Research Reports is to stimulate discussion and work in the hope of promoting and speeding resolution of identified issues. SAE EDGE™ Research Reports are not intended to resolve the issues they identify or close any topic to further scrutiny.

Edge Development Team

Dr. Nadia Vialas, *Head of Materials and Processes, Liebherr-Aerospace*

Alexander Altmann, *Head of AM R&T, Liebherr-Aerospace*

Claudia Jordan, *EOS GmbH Electro Optical Systems*

Chet Fuller, *Chief Commercial Officer of Norsk Titanium*

Robin Weston, *AM Marketing & Product Manager, Renishaw*

Jonathan Porter, *AM Business Development Manager, Renishaw*

Dr. Bhaskar Dutta, *DM3D Technology*

Dr. Joseph Flynn, *Assistant Professor (Lecturer), University of Bath*

John O. Milewski, *Los Alamos Nat. Lab. (retired), APEX3D LLC*

ISSN 2640-3536

Introduction

Additive Manufacturing (AM) processes form functional parts, of arbitrary form, from a three-dimensional (3D) computer model. This report will focus on processes which produce high-strength metal parts. Many different terms are used for AM processes developed by specific manufacturers. In January 2012, ASTM International produced a list of five standard categories for AM processes [1]. This list has since been adopted and expanded by the ISO. The seven categories now used by the ISO [2] are

- **Powder-bed fusion (PBF)**: Thermal energy (e.g., laser, electron beam (EB), or plasma arc) is focused to melt and fuse regions of powdered material, contained in a powder bed.
- **Directed energy deposition (DED)**: Thermal energy (e.g., laser, EB, or plasma arc) is focused to melt and fuse materials as they are being deposited. Wire feed or blown powder may be used as the feedstock.
- **Binder jetting (BJ)**: A liquid bonding agent is selectively deposited to join powder materials.
- **Material extrusion**: Material is selectively dispensed through a nozzle or orifice.
- **Material jetting (MJ)**: Droplets of build material are selectively deposited.
- **Sheet lamination**: Sheets of material are bonded to form a part.
- **Vat photopolymerization**: Liquid photopolymer in a vat is selectively cured by light-activated polymerization.

State of the Industry

The primary processes of interest for the production of high-strength metal aerospace components are PBF and DED. These processes involve a concentrated energy source together with the supply of metal or plastic feedstock. The feedstock is normally either a powder or a wire. The energy source melts the feedstock and substrate, forming a melt pool which solidifies layer by layer, to form the part. These processes may be used to produce high-strength metallic components and functional plastic components. PBF is used for smaller components produced to net shape, meaning that little or no final machining is required. DED is used for larger components and produces near-net-shaped parts, meaning that full machining is normally required to obtain acceptable surface finish.

BJ can be used to produce parts which are not fully dense and may require further processing to achieve fully dense parts. This can be a cost-effective way to achieve high-strength metal parts. Other methods are not typically considered for AM production of aerospace components. For example, material extrusion is normally referred to as 3D printing and often used for form-and-fit plastic models.

PBF and DED may be further classified according to their energy source (laser, EB, or plasma arc) or the feedstock used to deliver metal into the melt pool (powder or wire) [3], as shown in Table 1.

Manufactures often use their own trade names, sometimes creating the impression that their particular implementation is a distinctly different method of AM, as explained in Table 2. The terms used by manufacturers are also widely used with academic literature and, in some cases, are the original names used when the processes were first created. Most notably, Selective Laser Sintering was the first PBF method to be developed in the 1980s. Full melting of powder to produce fully dense parts was not yet possible. Selective Laser Melting (SLM) was later used to describe the modification of this process to allow for full melting of the powder feedstock. In this report the standard terminology is used where possible. It is also important to note, however, that hybrid AM and hybrid additive/subtractive machines are also emerging.

PBF builds a part layer by layer by scanning a laser or EB to fuse a profile in a powder bed. For each layer, the powder bed is incrementally moved downward and another layer of powder is added using a recoating blade or roller.

PBF-L is the most widely applied and perhaps the most evolved AM metal technology. A range of metal alloys suitable for aerospace components are available as well as high-quality polymers. Build volume is limited by the ability to maintain a consistent gas flow over the build area to ensure sufficient quality across the powder-bed area, and also by the requirement for the bed to be completely filled with powder.

TABLE 1. **Classification of the main AM processes.**

		Feedstock delivery	
		Powder-bed fusion (PBF)	Directed energy deposition (DED)
Energy source	Laser (L)	Laser powder-bed fusion (PBF-L)	Laser beam directed energy deposition (DED-L)
	Electron beam (EB)	Electron beam powder-bed fusion (PBF-EB)	Electron beam directed energy deposition (DED-EB)
	Gas metal arc (GMA)		Gas metal arc directed energy deposition (DED-GMA)
	Plasma arc (PA)		Plasma arc directed energy deposition (DED-PA)

TABLE 2. Terms used by specific organizations.

ASTM process	Feedstock	Organization	Manufacturer's name for process
PBF-L	Powder	EOS	Direct metal laser sintering
		3D Systems	Direct metal printing
		Concept Laser	LaserCUSING
		SLM Solutions	Selective laser melting (SLM)
		DMG Mori (previously Realizer)	
		Renishaw	
PBF-EB	Powder	Arcam AB	Electron beam melting
DED-L	Powder	DM3D	Direct metal deposition
		Optomec	Laser engineered net shape
		RPM Innovations	Laser deposition technology
DED-EB	Wire	Sciaky	Electron beam additive manufacturing (EBAM)
	Wire	NASA	Electron-beam freeform fabrication (EBF3)
DED-PA	Wire	Norsk Titanium	Rapid plasma deposition (RPD™)

This is heavy and expensive, and has safety implications. At present, the largest build volumes are approximately 400 mm × 400 mm × 800 mm. Deposition rates are approximately 5-20 cm^3/h for a single laser. Multi-laser systems are now able to achieve approximately 150 cm^3 in the fastest materials such as aluminum, and around 30 cm^3 to 50 cm^3 in nickel alloys. Lasers are mechanically steered using mirrors, meaning that multiple lasers are required to maintain multiple melt pools. PBF-L also achieves the best surface quality of any AM metal system and is related to powder particle size, typically ranging from 5 to 60 μm.

Sintering occurs when materials in close proximity and elevated temperature form bonds through a solid-state diffusion process. Complete melting, coalescence, or solidification is not required. When powder is sintered, it usually results in a part that is not fully dense and may require hot isostatic pressing (HIP) to fully consolidate the material. Early powder-bed processes sintered rather than fused powders. Coated metal parts were sintered and then infiltrated with a lower melting point metal, such as copper or bronze. Metal fusion occurs when sufficient melting occurs to overcome surface conditions so that the material coalesces

FIGURE 1. **Metal is shaped through the use of a laser sintering machine.**

MarinaGrigorivna/Shutterstock.com

and mixes before cooling and solidifying. Fusion can produce fully dense parts with high-quality microstructures. The latest metal-sintering processes are liquid-phase sintering, in which some alloying components melt. This blurs the line between sintering and fusion, and can achieve near fully dense parts that do not require infiltration [4].

PBF-EB can typically use the same alloys as PBF-L although materials must be conductive, so it is not suitable for polymers. Processing is typically carried out in a vacuum, and metallic powders are heated to approximately 700°C to reduce residual stress and distortion. This can also partially pre-sinter the powder surrounding the part, reducing the need for support structures and simplifying the stacking of multiple parts within a single build. Build volumes are currently limited to approximately 350 × 350 × 380 mm. The ability to rapidly steer the EB using magnetic field deflection enables a single beam to maintain multiple melt pools. Build rates today can be up to 80 cm^3/h. The surface finish is not as good as for PBF-L, primarily due to the requirement for larger powder sizes of approximately 45-105 μm. Arcam AB is so far the only commercial manufacturer of PBF-EB.

DED involves delivering the powder or wire into the focal spot or molten pool created by a localized directed energy source such as a laser, EB, or plasma arc. A key advantage is that the build volume is not limited by the fixed size constraint of a powder bed. It is also typically able to achieve higher deposition rates, when compared to PBF, but at the expense of surface finish and greater potential for distortion. DED is much more likely to require postprocess machining, as well as stress relief, than PBF.

Directed energy deposition using metal arc weld (DED-GMA) and plasma arc (DED-PA) are based on well-established welding methods. They are able to rapidly and cost effectively produce large, complex, near-net-shaped components. Robots may be used to manipulate the welding head as it deposits material, resulting in very large build volumes at relatively lower costs. A wide range of commercially available weld wire alloys can be used, further reducing cost. The large melt pool results in weld bead-shaped deposits which are normally machined to achieve the required dimensional accuracy and surface finish. The process may need to be completely contained within an inert gas chamber due to extensive heat buildup when depositing reactive or refectory alloys. Distortion and residual stress relief normally require postprocess heat treatments [5].

DED-L, based on 3D laser cladding, is used to repair and deposit thin features onto base parts, including complex surfaces. Powder feedstock can be deposited at up to 500 cm^3/h and have less demanding size and shape requirements compared to PBF. Build envelopes of up to 1500 × 900 × 900 mm are available. Issues with distortion and residual stress may require heat treatments and HIP (Figure 2).

DED-EB is a near-net-shaped process based on EB welding and weld cladding technology. It allows very large build envelopes of approximately 1800 × 1200 × 800 mm. Commercial weld wire is deposited at extremely high rates of 700-4000 cm^3/h. The large melt pool results in weld bead-shaped deposits which require additional machining. Distortion and residual stress can also be significant necessitating further processing, such as heat treatments.

BJ can be used to bond metal powders such as stainless steel or tungsten. The resulting "green" part will not be fully dense. Mechanical properties may be improved by infiltrating with a lower melting-point matrix metal such as bronze, although the properties of an infiltrated sintered metal cannot compete with wrought alloys. In some cases

FIGURE 2. Laser sintering machine building up layers of a metal component.

sspopov/Shutterstock.com

HIP is able to densify components without infiltration at the cost of significant shrinkage requiring compensation of the design and build models.

BJ is most commonly used to produce molds for casting. Material extrusion, or "3D printing," may also be used to produce wax or polymer patterns for investment casting.

Sheet lamination of metals typically uses ultrasound to solid-state bond sheet metal. Machining is then required to remove the part from the base plate and the bulk of unfused layers. Dissimilar metals can be bonded in this way, creating a functionally graded part. Cold spray technology can be used to consolidate powders onto a substrate to deposit a 3D shape build.

Hybrid machines have been produced which integrate a DED-L with a powder feed head into a Computer Numerically Controlled (CNC) machine tool [6]. This allows complex features to be added to simple base shapes. It can be useful for small and complex parts produced from hard to machine materials, or for larger parts that would require high stock removal volume. Both AM and machining can be carried out with a single setup.

Advantages of AM Parts

A major driver for aerospace components is weight reduction. Depreciation accounts for approximately 12% of airlines' operating costs, compared with 36% for fuel [7]. Customers are, therefore, more interested in performance than cost. AM components produced using PBF and DED have more refined grain structures than cast components, comparable with wrought components.

Another significant performance benefit of AM components is the ability to precisely control internal geometry to create cooling channels and heat exchangers with features such as turbulators which increase heat transfer.

Despite aircraft operators' relative insensitivity to price, once orders have been secured, any reduction in manufacturing cost translates into profit for the aerospace manufacturer. In general, AM is nominally more expensive than a traditional manufacturing process, such as machining or casting. However, there are potential cost savings from component consolidation, reduction of process steps, and reduced material waste.

The ability of AM to produce complex shapes often enables many components to be consolidated into a single part. Part reduction can greatly simplify the many operations required to move from a prototype to a certified part and by reducing the bill of materials. For example, equipment calibration, operator certification, development of standards, documentation, records keeping, and formal specifications and qualified materials, operators and suppliers are all reduced. Component certification costs can be many times more than design, so component consolidation can result in significant cost savings.

Reducing the number of processes required to produce a component also has similar advantages in terms of reducing certification overheads. Although a component may be more expensive to manufacture using AM rather than a conventional process such as machining, the total cost of the certified part may be lower.

AM can significantly reduce material waste when compared to machining from billet. This is of particular importance for highly specialized alloys such as nickel-based alloys or titanium. Machined components, with buy-to-fly ratios as high as 40:1 might see this reduced to 4:1 using AM. It is not possible to completely eliminate waste due to the need for support structures, combined with final machining for DED or loss of powder quality in PBF.

AM Parts in Aerospace Production

Despite remaining challenges, AM is being used to produce aerospace components that are already flying. A wide variety of alloy aero-engine components are being produced using PBF as well as brackets, heat exchangers, and now even structural airframe components.

Perhaps the most well-known example of an AM aerospace component is the General Electric Aero LEAP fuel nozzle. This consolidates 18 components into one, with complex internal channels [8]. GE has been producing the component using PBF-L since 2015 and is on track to produce over 30,000 engines a year by 2020 [9], with this component already flying on hundreds of aircraft. GE has gone even further with their Advanced Turboprop engine which will power the new Cessna Denali. Over 800 conventionally manufactured parts were consolidated into 12 PBF-L parts.

Lockheed Martin has used Sciaky's EBAM process (DED-EB with wire feed) to produce titanium blanks from which propulsion tanks are machined, removing the need for punch-and-die tooling to form sheet titanium [10].

Airbus started producing a double-walled pipe elbow for the A400M fuel system in 2016. They are now flying thousands of different types of plastic brackets, clips, and other devices for securing cables, wires, and hoses. Many are made from ULTEM 9085 on a Stratasys FDM, including 2,700 parts for the A350 XWB. Materialise is the first serial supplier of these parts. Airbus is also flying serially produced Ti-6Al-4V parts produced by subsidiary Premium AEROTEC and has used Sciaky DED to produce Ti parts [11].

Establishing the airworthiness of aircraft components is the responsibility of the Federal Aviation Administration (FAA). An airworthy aircraft must "conform to its type design" meaning that all parts are produced to the parameters specified in the type certificate, and it must "be in a condition for safe operation," taking account of corrosion, cracks, etc. [12]. Obtaining certification involves very significant testing of components produced using processes with fully documented process parameters. Subsequent production of these components must demonstrate that the parameters remain within tolerance.

Norsk Titanium uses wire-fed plasma arc directed energy deposition (PA-DED) to produce titanium alloy preforms, which are machined into structural airframe components. This is the first process to receive FAA certification for additively manufactured structural airframe components. Norsk is a tier-1 supplier to Boeing, enabled by the adoption of SAE standards [13, 14] to demonstrate that traceable processes are used. Boeing is also flying tens of thousands of polymer PBF-L parts on at least 16 types of commercial and military aircraft. Parts are also produced from PEKK on a Stratasys PBF-L machine.

Unsettled Topics

A number of generic challenges and unsettled topics relate to the shift to AM production in general, regardless of the specific method employed. These challenges are

- Design optimization
- 3D geometry representation and preprocessing
- Understanding and modelling fundamentals of process
- Process sensing and control - access to process parameters
- Establishing material properties
- Defect detection

These generic challenges are discussed in the below sections.

Design Optimization

The geometrical freedom afforded by AM processes makes design automation an attractive possibility. A widely publicized application is structural topology optimization. Here, an iterative simulation loop (finite element analysis, FEA) removes material from the component, while adhering to constraints imposed by load cases and material properties. The organic forms that arise are challenging to produce with traditional manufacturing processes due to complex tool paths and tool access, costly tooling or fixtures, and excessive buy-to-fly ratios (Figure 3).

At present, material properties arising in AM processes still have larger uncertainty than most traditional processes. Uncertainty comes from many sources, including

- The inherent anisotropy due to a layer-wise manufacturing process
- The complex thermal history that develops as a component is built
- Variation in the feedstock
- Variation in surface roughness
- Flaws such as micro-cracks, porosity, and partially bonded particulate
- Position of the component on the build plate due to varying gas flow across the build chamber

Structural FEA only represents one discipline within computer-aided engineering. Topology optimization based on physics, such as electromagnetics, heat transfer, and fluid dynamics, have all been explored within academia [15, 16, 17].

FIGURE 3. **Topology optimized component manufactured through AM techniques.**

MarinaGrigorivna/Shutterstock.com

However, none of these have reached the prevalence of structural applications. Structural applications have an intuitive connection between the simulation objective and the placement of material, i.e., low stress leads to less material. However, linking a particular fluid-flow characteristic to the placement of material is less intuitive and requires enormous computation.

Recommendations In the immediate future, topology optimization should include uncertainties arising from variable material properties and the variable nature of the AM process [18]. Improved data on materials properties is first required, discussed in more detail below. This must identify how fracture toughness and fatigue are affected by defects in thin-walled and slender sections. There is also a need for a better understanding of process variables with particular reference to defects and flaws.

As an ongoing concern, the community should consider the wider application of design automation (i.e., simulation in the loop) based on different branches of physics. Many recent success stories within the AM community relate to heat transfer and fluid flow [19, 20]. This affects products such as heat exchangers, hydraulic manifolds, and the general application of conformal cooling channels. In these applications, the benefits of lightweighting are combined with improvements in pressure drop or heat transfer effectiveness. This combined benefit arguably exceeds the benefits of lightweighting in isolation.

3D Geometry Representation and Preprocessing

A "good" piece of design for AM can be linked to having a large surface area-to-volume ratio (SAVR). This implies that the material has been used sparingly, the topology has become more complex (e.g., lattices), internal surfaces are being exploited (e.g., cooling channels), etc. However, herein lies a problem. As SAVR increases, so does the computational overhead associated with rendering, slicing, supporting structure generation, and the creation of machine code for the AM process.

Tessellated file formats dominate the world of AM. The true geometry of a component is represented by a surface mesh of tessellated elements, usually planar triangles. Popular incarnations include the .STL file and the .OBJ file. This representation will only approximate any curved surface. As the SAVR of the object increases, or areas of high curvature emerge, the number of triangular facets rapidly increases. For context, the .STL file of an intricate lattice can easily extend beyond 10 GB. Loading this file into RAM and then manipulating the geometry involves huge levels of computation, and can halt the design process. Furthermore, meshes are prone to holes, triangle intersections, and other such defects [21]. Healing these becomes cumbersome with increasing mesh size.

Recent developments have tried to extend the capability of tessellated file formats by permitting mesh elements with curvature (e.g., the .AMF [22]). This means that considerably fewer elements are required to achieve the same level of accuracy in the component representation. However, removes one of the few benefits of planar facets, which have easy to implement and efficient algorithms for editing, healing, and creating machine code. Curved elements remove this advantage, which simply relocates the computational bottleneck. However, additional benefits are afforded by newer file formats, such as AMF and 3MF [23], in particular the ability to specify colors and materials, as well as being backwards compatible with the .STL format (Figure 4).

FIGURE 4. Example of the type of complex geometry and color information that the new file formats are able to represent.

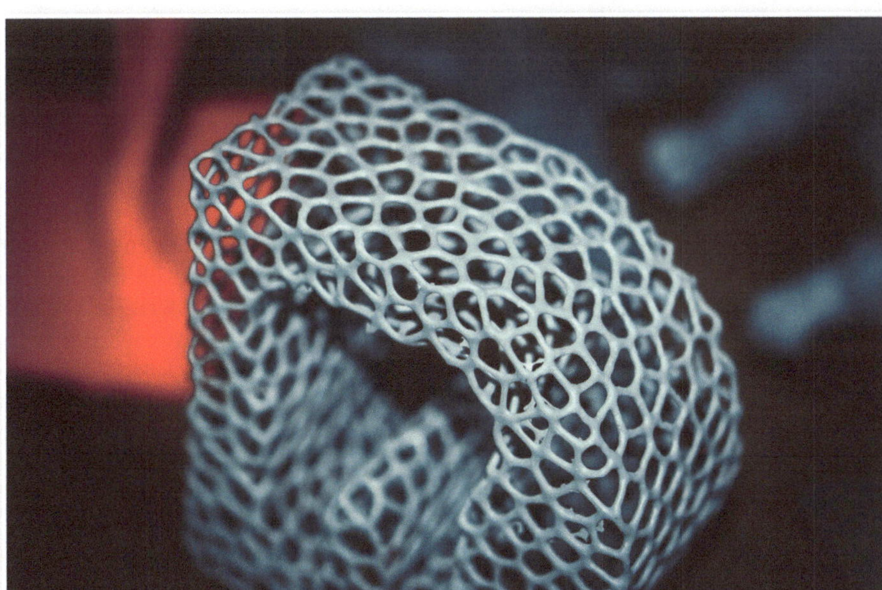

MarinaGrigorivna/Shutterstock.com

Within mechanical engineering, the computer-aided design (CAD) market is dominated by four vendors, and all software packages are driven by parametric curves and surfaces. This type of geometry gives rise to the familiar formats exported, CATIA, SolidWorks, Creo, NX, and Inventor. Ideally, AM processes would be fully integrated with this format, but unfortunately most AM machines do not process geometry this way. Though highly developed and powerful in many ways, parametric surfaces are also ill-suited to process very complex geometries. For example, embedding a lattice with many thousands of elements will bring most workstations to a standstill without very careful management. In addition, computing intersections for features such as fillet radii often leads to significant manual intervention and prohibitive levels of computation [24].

Recent products in the AM software market are now moving in a new direction, namely, implicit geometry. These representations are highly efficient in terms of rendering, slicing, and complex geometry operations. They are particularly well suited to lattice-like geometry and are the ideal representation for triply-periodic minimal surfaces, which are a growing concern in terms of lightweighting, impact attenuation, heat transfer, and more. This geometry representation is very promising, but significant additional work is needed to integrate it into the conventional workflow of design engineers. Immediate concerns relate to working with assemblies, geometric dimensioning and tolerancing, and the ability to run computer-aided manufacturing software for downstream processes.

Recommendations AM users should consider the entire process chain (design, print, postprocess, inspection, certification) before committing to a workflow or geometry representation. Reliance on traditional computer-aided design (CAD; parametric surfaces) is easily the least disruptive but is limiting for highly complex geometry. Tessellated mesh formats (STL, OBJ, AMF, and 3MF) still dominate the market. However, these are also prohibitive with very complex geometries and will usually require mesh-healing software before printing. Implicit geometry is new to the market but holds great promise. However, until wider design considerations are factored into this format, it places severe constraints upon the design and makes the workflow. In the short term, many of these constraints could be avoided through a robust conversion between implicit geometry and traditional CAD (parametric surfaces).

Understanding and Modelling Material Deposition

Achieving stable processes which produce consistent components require an understanding of how process variables affect the properties of deposited material and the component geometry. Process input variables affect both the grain structure of deposited material and the formation of defects such as porosity and delamination [25].

Ideally, a first principles physics-based model would be validated using a full-factorial design of experiments (DOE). It would then be possible to predict, with confidence, material properties for any combination of process parameters. However, in practice AM processes have a large number of parameters, meaning that such a complete validation is not possible. In total there may be around 30 parameters for an AM process; even with only two-factor levels, a full factorial design would therefore require over one billion coupons to be produced. Models must, therefore, be used for process optimization, to determine the most critical parameters and to establish permissible ranges for these parameters.

DOE may then be carried out for the primary key process parameters (KPPs), which are believed to have the most impact on the process. For example, Norsk Titanium has completed a DOE with 6 KPPs and 45 KPP combinations, including all combinations that could produce stable beads, ignoring those that would obviously fail dimensional inspection. Parts were tested for dimensional variation, hardness variation, microstructural variation, tensile variation, and full chemical analysis. This was sufficient to gain FAA qualification for Boeing parts.

The fusion of metal in AM processes is very similar to metal fusion in welding processes, with the same sources of concentrated heat being used. Laser and EB welding have been used for over 50 years while arc welding has been used for more than 100 years. Understanding gained from these processes serves as the basis for understanding AM processes [26, 27, 28, 29]. It is also relevant that welded aerospace components have been certified, demonstrating a route to certification for AM components.

Some important fundamental considerations include the way the energy source interacts with the material, the fluid flow, and dynamics of the melt pool. When energy is directed at a metallic surface, some of the energy never reaches the surface and is lost to the environment, some is reflected as it hits the surface, and some energy is absorbed but then radiated away. Reflection and radiation depend on wavelength, material, and temperature. For example, molten metal absorbs much more laser energy than solid metal. Fluid flow within the melt pool and vaporized metal flow above the pool are highly dynamic and difficult to control. Flow within the melt pool breaks up surface oxide layers between powder particles and filler wires, and assists fusion into the previously deposited layers. Vaporized metal above the molten pool can disturb the process in unpredictable ways, alter alloy chemistry, and condense on machine components, interfering with correct function. Vaporized metal above the melt pool may absorb laser energy preventing it from reaching the melt pool and causing the cloud of vapor and gas to expand rapidly. The resulting pressure can depress the melt pool, creating a keyhole that can extend deeply into the metal leading to defects, such as porosity, spatter, or balling, and entrapped voids. Sudden transitions in the melting mode can be caused by small changes in laser power, travel speed, focal spot size, and other process disturbances. Electric arc characteristics are strongly dependent on the direction of current flow (AC, DC+, DC–), shield gases such as nitrogen or argon, and feedstock properties [30].

Simulation of laser, or EB, material interactions is challenging. FEA models of these effects require a high level of mesh refinement around the melt pool, thin features, and submodels of powder boundaries. The mesh must extend through the powder layer, through the fused deposit layer to a depth of three or more layers where full fusion occurs. This is combined with a need for high temporal resolution required to model dynamics effects. This can be computationally challenging. Simulations are also dependent on accurate material property data over the wide range of temperatures from ambient to vaporization, the available data is often not of sufficient accuracy to produce realistic simulations.

An alternative to first principles physics-based analysis is to create data-driven models by fitting low-order functions to map input to output conditions. Simple models may be referred to as regression while more complex models employ deep learning. Such an approach may be beneficial, particularly when optimizing process parameter sets and determining the sensitivity of output variables to variations in input variables. However, the number of process inputs for AM processes is substantially higher than for conventional manufacturing processes. It is not possible to map all the process inputs, and it is important that the standard of proof applied to a new process is similar to that applied to existing processes, thus supporting the adoption of new technology.

There is ongoing work to identify the primary KPPs, a subset of parameters that have the greatest impact on the part properties to increase confidence in and the reliability of the process. Experiments have been carried out for proprietary processes with thousands of coupons leading to "frozen" parameter sets that yield material properties that are stable enough for certified processes.

Recommendations Improved knowledge of the fundamental processes involved in AM is required, leading to an understanding of the relationship between process variables and defect generation, microstructural evolution during solidification [31], and solid-state transformations upon cooling. This will enable the optimization and control of process parameters.

DED-GMA is relatively well understood due to its similarity with arc welding processes. The extension of weld models to DED-GMA must account for deposition paths, distortion, and shrinkage compensation design considerations that are much more complex than most weld applications. As such, the thermal and mechanical history of any material deposited into a complex shape will evolve microstructures and stresses that can vary throughout a component. The scale of critical features such as flaws and defects is generally within the detectable range of current nondestructive evaluation techniques.

PBF processes are less well understood and much more difficult to model. Standards for powder feedstock are being developed. There are varieties of vendor-dependent powder production methods for a given alloy that produce multiple chemistries and powder morphology. Process sensing and controls are less evolved than those for arc weld-based heat sources. As an example, a sensor to characterize the spatial energy profile of the laser impinging on a powder bed at any location within the powder bed does not yet exist. This data is critical to a first principles physics-based model. Many of the other first-order inputs such as travel speed, laser power, track overlap, and focal conditions, at every point along the deposition path, are either not known or have been hidden from the user as proprietary intellectual property. The scale of critical features such as flaws is in some cases beyond detection, particularly in complex components. Improved material and process characterization is needed to improve inputs to mechanistic models. The degree to which standards improve our understanding and control of the critical process inputs is yet to be determined. Technology based upon big data collection, machine learning, and digital twin technology *combined* with first principles models may offer a path forward to prediction and control of these complex processes [32].

Key areas where improved modelling would increase understanding are the build-up of residual stress, modelling of efficient heat dissipation to enable faster builds, and modelling how parameters affect microstructure.

Process Control

Once the fundamentals of material deposition are understood, optimized process parameters and parameter tolerances should be identified for specific machines and processes. Issues regarding material properties, product consistency, process repeatability (e.g., same machine, different day or moving from one batch of powder to another), and process transportability (different machine, at a different location, using the same parameters) need to be fully addressed to gain the confidence required for material and process standardization and certification when used in critical applications.

Key process variables include the melt pool size, powder layer thickness, heat source power, focus condition, travel velocity of the melt pool, and scan line offset. These are critical for complete fusion into adjacent layers and penetration into previous layers. Powder parameters include chemistry, particle size distribution, and shape.

Standard parameter sets supplied by vendors may be protected by encryption techniques that prevent the user from seeing the actual parameters or modifying them to suit in-house development needs. Some equipment vendors are now providing open architecture. However, with insufficient modelling of material deposition, selecting parameter values is challenging.

Power and travel speed may be referred to as primary process input variables, these may be changed and controlled during processing either by the operator or by an automated control. Secondary process-input variables are normally held constant during the process development cycle. Examples include inert shielding gas or feedstock type.

Primary process-output variables can be observed by inspecting the deposit, for example, melt bead profile, depth of penetration, surface condition, and deposit density. Secondary output variables are typically harder to observe flaws and defects, residual stress, or thermal softening.

One challenge for AM process control is understanding whether builds can be restarted. For example, the process may have to be stopped during the build, or a defect identified after the build may need repair.

Many AM systems already offer real-time melt pool sensing and control. Sensors include digital cameras, spatially resolved thermography and pyrometry, and laser profiling or EB diagnostics. Further work is needed to optimize those capabilities for gathering of sufficient data for in-process quality assurance.

SAE's Aerospace Materials Specification (AMS) standards are the state of the art for AM process control. AMS7003 is a generic standard that defines the requirements for traceable PBF processes [33]. The AMS7005 standard is a major milestone which establishes process controls for the repeatable production of preforms by PA-DED. It specifies that values, tolerances, and measurement frequency must be established for at least 28 key process variables which it identifies, in detail. It allows for proprietary variables, which may be considered as commercially sensitive by a producer, to be assigned a code and documented internally by the producer [13] (Table 3).

Recommendations Standards such as AMS7003 and AMS7005 should be used as the framework for greater standardization of parameter sets. Ideally, proprietary parameter sets would be replaced by industry standard parameter sets, analogous to standard feeds used in machining. However, it is likely that increased process complexity and differences between machines produced by different suppliers may make this impossible. Quite possibly each manufacturer's proprietary approach will need its own qualified and approved parameter set, as is already the case with die-forging processes.

Establishing Material Properties

Initial studies indicate that the tensile properties and hardness of AM metal components are as good as cast or wrought parts [34, 35, 36]. However, less than full-density materials may have compromised fracture toughness and fatigue resistance since porosity or slight delamination can act as crack initiation sites. Further work is therefore required to fully characterize the properties of these materials together with their dependence on related process parameters and observable defects.

Large material property databases have been compiled for conventional metals over the past 100 years [37, 38]. These provide detailed performance data related to specific operational conditions in applications such as aviation, spaceflight, and automotive. By comparison, the publically available data available for AM materials does not provide a high level of confidence in the airworthiness of AM components.

An extensive review was carried out in 2016 and compiled the available data for a range of alloys and PBF and DED processes. This found that although initial tensile data is very promising, there is very little published work for low-cycle fatigue, fatigue crack growth, fracture toughness, impact, creep, creep fatigue, multiaxial testing, or environmental effects. The variability of properties such as microstructure, residual stress, and defects was also found to be a concern [39].

Melt back penetration from the bead being deposited into a previous layer can solidify and continue the previous layers' grain orientation, known as epitaxial growth. It can result in stress concentrations, warping, and other anisotropic properties. AM processes may avoid this by changing the deposition path between layers. There are also attempts to model and utilize these effects to control the microstructure, an approach known as microstructural engineering [40].

The first independent standard for an AM material is AMS7004, released by SAE International in early 2019. This covers preforms of between 40 mm and 254 mm thickness using a PA-DED AM process on a Ti-6Al-4V substrate, subjected to post-deposition stress-relief heat treatment. It provides detailed specification of the key process variables that must be controlled for the material properties to be known. It gives tensile stress, yield stress, and elastic modulus for the material [14].

Although public standards and databases do not contain sufficient fatigue and fracture toughness data, internal data related to proprietary processes is available and enables critical structural components to be certified. Thousands of coupons are produced and tested to determine frozen parameter sets which provide the required confidence and process tolerances.

Recommendations Further work is required to catalogue material properties for a range of alloys and process parameter sets. Particular focus is needed to determine materials' low-cycle fatigue, fatigue crack growth, fracture toughness, impact, creep, creep fatigue, multiaxial testing, or environmental effects. The variability of these properties as well as their sensitivity to variations in process parameters also needs to be determined. Such an effort requires industry-wide collaboration, potentially coordinated by the government.

Where certified AM parts are already being used in critical areas, the material properties have already been established, but this may be proprietary information. The machine-specific nature of AM processes means that, for the foreseeable future, material properties will be given for a specific proprietary process. This does not have to mean, however, that the data remains proprietary and the inclusion of detailed material properties for proprietary AM processes in databases, such as the Metallic Materials Properties Development and Standardization (MMPDS).

TABLE 3. Key process variables for PA-DED, adapted from SAE AMS7005 [13].

Variable	Description	Impact to process
Interpass temperature	Temperature of the workpiece before depositing a new bead	Impacts solidification and cooling rates and, therefore, microstructure
Interpass time	Time between ending one bead and starting the next bead	Impacts thermal processing parameters required for deposition
Preheat arc current, type, and waveform	The preheat arc is transferred from the electrode to the workpiece	Impacts the temperature of the areas ahead and underneath the actual deposit and can impact the melt pool conditions
Preheat torch plasma gas flow rate	The preheat plasma gas supplies continuously inert gas to be ionized near the electrode and form the plasma gas	Impacts the capability of the preheat function
Preheat conditions	The thermal condition of the workpiece prior to start of deposition	The thermal condition can impact material quality and workpiece distortion
Main arc current, type, and waveform	The main arc is transferred from the electrode to the wire	The main arc is the main contributor to the wire melting
Melter torch plasma gas flow rate	The melter plasma gas supplies continuously inert gas to be ionized near the electrode and form the plasma gas	The melter gas flow rate impacts the weld bead and fusion characteristics
Standoff (arc power) control	Distance between the melter torch and the workpiece	The standoff distance determines the arc resistance and thereby impacts the arc power
Wire feed speed	The wire is fed to the melting zone at an established speed	The wire feed speed defines and impacts the melting rate and the weld bead profile
Traverse speed	The rate of relative movement between the torch system and the workpiece	The traverse speed impacts the weld bead width and height. Impacts material quality and rendering accuracy
Plasma transferred arc (PTA) current, type, and waveform	The PTA is transferred from the electrode to the workpiece	The PTA provides an additional heat source for the workpiece, and for the establishment of the weld pool
Preheat and melter torch shielding gas flow rate	A gas used to produce a protective atmosphere	The shielding gas flow rate can impact the weld pool by providing additional gas to secure an inert shielding atmosphere
Plasma and shielding gas composition	Chemical composition and purity of gas input into plasma and shielding gas system	Inappropriate composition can impact chemistry of the deposited preform and material quality
Build chamber gas composition	The gas composition inside the build chamber during the deposition	Gas composition can impact material quality
Build chamber moisture content	The moisture content inside the build chamber during the deposition	Moisture content can impact the oxygen pickup in the material and affect the mechanical performance of the material produced
Build chamber oxygen content	Oxygen content in the build chamber during the deposition	High oxygen content can lead to oxidation of titanium alloy. Supply gas should be measured and monitored
Plasma torch specification	The device that generates the plasma transferred arc	Traceability of critical equipment for the deposition process
Electrodes and consumables specification	Material, physical characteristics and method of manufacture for the electrode and consumables	Impact the metal transfer characteristic. Impact the interaction with the workpiece
Electrodes and consumables erosion rate	Variation in mass, geometry, and surface finish of the electrodes and consumables	Impact the metal transfer characteristic. Impact the interaction with the workpiece and the deposited material quality
Feedstock specification	Material specification for the wire and substrate to be used for this application that defines the physical characteristics, and method of manufacture	Statistical process control (SPC) can be applied against the physical characteristics defined within the feedstock specification. Method of manufacture is discrete and not required for SPC
Feedstock diameter	The diameter of the feedstock allowed for the process	The feedstock diameter can impact the metal transfer characteristics and the integrity of material produced
Substrate specification	The material, dimensions, tolerances, and surface finish of the substrate	Could impact geometry through distortion which would be identified in geometric inspection. Incompatible material and surface finish could negatively impact the process

TABLE 3. (Continued) Key process variables for PA-DED, adapted from SAE AMS7005 [13].

Variable	Description	Impact to process
Deposition strategy	Bead deposition order and direction, bead width and height, bead overlap, cooling rates, energy input	Impacts the material quality and productivity
Feedstock manufacturer	The company producing the wire and substrate	Variation in wire and substrate between producers can impact the integrity of the material produced
PA-DED machine manufacturer, model, and serial number	The overall configuration of the PA-DED system	Hardware configuration
Process interruption	The allowable break in the continuity of the deposition process as substantiated and approved by the chief executive officer	A delay in the deposition process can have an impact of the thermal profile of the deposited material and the resulting microstructure produced
Integrated control system	The hardware, software, firmware, programmable control modules that allow operation and control of the system	Integrated architecture operation required for process to be performed
Power of each plasma arc	The commanded power and the delivered power at the workpiece surface	Impacts the material quality

Defect Detection

AM part inspection involves dimensional measurement, surface measurement, and nondestructive tests (NDT) to identify internal defects. Flaws and defects in AM metal parts include the lack of fusion, porosity, voids, cracking, oxidation, discoloration, distortion, irregular surface profile, or surface stair stepping. Cold lap is a lack of fusion caused by an oxide layer or thin film. Slumping is a dimensional defect normally due to insufficient support structures during solidification. Rounding of edges may occur due to surface tension in the melt pool.

Dimensional inspection of prismatic features is well developed with coordinate measurement machines (CMM) able to provide micrometer-level uncertainties and automated rapid inspection. Inspection of complex forms, such as generated by topological optimization, may be time consuming, using a CMM and noncontact scanning may therefore be considered. However, demonstrating traceability for noncontact scanning measurements remains challenging and uncertainties are considerably higher.

During AM, only the last layer fused can be observed. However, when further layers are deposited over the top, the current layer may re-melt causing deeper penetration and the closure of apparent defects. The definitive tests for porosity and delamination are destructive, involving sectioning test pieces and observing their internal structure under a microscope. In high-volume production, such methods may be used to sample parts and validate other quality-control methods, including controlling the process input variables. There is, however, often a requirement to detect flaws and defects in parts using NDT. Methods include visual inspection, dimensional inspection, dye-penetrant inspection, leak testing, radiographic, computed tomography, and proof testing [41].

Porosity is a common defect in melted and fused material, it can be caused by gases such as hydrogen entrapped within the melt pool, keyhole collapse, or fusion defects. It is important to differentiate between gas porosity and other forms of voids such as the lack of fusion voids as they have different shapes leading to different crack propagation behaviors and different origins, requiring different methods for avoidance and control.

Radiographic testing uses X-rays or gamma rays to view the internal structure of parts. It is widely employed to inspect welds in pipelines, pressure vessels, and other critical applications. It can sometimes detect the lack of fusion defects. Computed tomography (CT) uses a number of sectional images, obtained by X-ray or magnetic resonance imaging, to view the three-dimensional internal structure. The lack of fusions is especially challenging for CT due to their small volume. Accuracy depends on factors like the material being used (density), the penetrations lengths, and the geometrical complexity of the application being manufactured. This restricts the suitability of conventional NDT to AM. Cost is also an issue for CT inspection.

Surface roughness of as deposited AM parts is an issue for many NDT techniques, including dye-penetrant inspection and CT. This means that surfaces often need to be machined before NDT can reliably be performed.

Eddy-current testing is used to detect surface and subsurface flaws, such as cracks and pits, in conductive materials using electromagnetic induction. It can be sensitive to surface and near-surface conditions and material type. Laser ultrasound has also shown some potential as a possible method of detecting porosity [42]. Immersion ultrasonic testing (UT) and phased array UT are also being investigated for AM parts.

Due to the high cost and inability to detect all defects of NDT methods, there is significant interest in the use of in-process monitoring.

Recommendations Dimensional tolerances, both for form and fit and for stress analysis, should be specified with consideration of how they will be verified. It may be assumed that relatively simple geometry will be verified using a CMM with typical uncertainties of approximately 5 μm, meaning that tolerances should be no smaller than ±30 μm. For topology optimized geometry, noncontact measurement may have an uncertainty of 0.1 mm, meaning tolerances of ±0.5 mm would be more realistic.

If NDT techniques can be enhanced to allow reliable inspection on as-built surfaces, that will immensely help the certification process and reduce the costs of manufacturing. The next step may be development of efficient and reliable in situ inspection techniques. Since AM is a layer-by-layer process, it allows in situ inspection opportunity in between the layers. This can be very significant particularly in the initial development and certifications stages.

Unsettled Topics Concerning PBF Processes

Amortized Machine Cost

The amortized machine cost is essentially the cost of machine ownership per kilogram of part produced. It is therefore a function of the capital cost of machine purchase, machine maintenance costs, machine utilization, and deposition rate.

Although the prices have dropped considerably, high-powered lasers used for AM systems can still cost hundreds of thousands of dollars. EB-based systems are also a major expense. Multiple laser beams can increase the speed of the powder fusion portion of a build cycle, but it cannot increase the speed of the powder recoat cycle or other system pauses. Although currently only available for polymer PBF machines, the EOS LaserProFusion system dramatically increases deposition rates by using an array of millions of laser diodes instead of a single scanning laser. This enables whole layers to be fused in a single operation, showing the potential for PBF processes to be accelerated [43]. The rapid magnetic steering of EBs enables these systems to maintain multiple melt pools simultaneously, thereby increasing process efficiency.

Larger build volumes not only require more material for the actual part but also generate more material requiring reuse or recycling. Powder-bed system raw material requirements will scale directly with the build volume. Powders can represent significant costs in PBF systems. Although material utilization can be high by sieving and recycling unmelted powder, large quantities must still be purchased and properly stored and handled. It is not uncommon for parts weighing just several kilograms to require hundreds of kilograms of powder in the bed. Scaling to very large parts quickly becomes uneconomic.

Recommendations Continued development of machines, together with greater machine production volumes, is expected to lead to lower machine costs and faster deposition rates. It may be possible to incrementally increase deposition rates by a factor of 10 or more with novel. However, there is a trade-off between deposition rate, distortion, and other scaling issues.

In the short term, optimized process parameters and the use of multiple lasers within a single machine are increasing deposition rates. In the longer term, metal PBF may see a step change in deposition rates with arrays of laser diodes fusing whole layers, as is currently the state of the art for polymer PBF.

Powder Cost and Availability

AM metal machines rely on very high-quality metal powders and wire that may be difficult to obtain for specialty alloys. Powder prices are not published but range from \$20 to \$250 per kilogram depending on the value of alloys; value of precursors such as processing chemicals, scrap metal, and metal wire; size tolerance; and process costs [11]. Prices for aerospace PBF are at the upper end of the range, making material costs considerably higher than the comparable feedstock used in conventional manufacturing processes.

Materials used in aerospace AM include aluminum alloys, stainless steel alloys, titanium alloys, nickel-based alloys, cobalt-chrome alloys, and also refractory metals such as tungsten, molybdenum, rhenium, niobium, and tantalum.

Industrial production of powdered metals has existed for some time with applications including sintered gears and weld filler. Commercially available powder comes in many sizes and shapes and is produced using a wide range of processes. However, the requirements for metal powders used in PBF are more demanding with respect to size, shape, and chemistry. These are critical process variables which affect the repeatability of the process. For example, powder produced by water atomization may be angular or irregular in shape, with sizes ranging widely from less than 1 μm to over 100 μm. This makes these cheaper powders unsuitable for PBF. Good sphericity and sizes between 10 and 100 μm are typically required for PBF so that powder can be spread into even layers. Powders produced by precipitation or crushing may require re-melting and sieving to achieve geometrical specifications. Alloy composition may also require adjustment to compensate for selective vaporization during the process.

High-quality titanium alloy Ti-6-4 powder is produced using hydride-dihydride, gas atomization, plasma atomization, and the plasma rotating electrode processes. Powder sieving may be required to remove out of tolerance particles. The hydride-dihydride process uses a hydrogen reaction to form and extract a metal hydride compound, which is later reacted back to metallic form. An additional plasma melting process may be needed to produce spherically shaped particles. The gas atomization process involves the metal alloy being induction- or furnace-melted and atomized with a stream of gas, typically argon, collected, and sieved into the desired range of sizes. The gas atomization process can produce high-purity, spherical powders. However, gas porosity entrapment may lead to gas porosity within AM components. Plasma atomization can produce very spherical uniform powders from wire feedstock.

Improved purity, cleanliness, and geometrical uniformity of powders will lead to improved AM process repeatability, a requirement for certification. At the same time, the cost of powders must be reduced, bringing them closer to the price of conventional feedstocks. More economical and environmentally friendly ways to extract metals from ores could significantly reduce the cost of titanium [44, 45]. Very recently, there have been significant investments made to improve the supply of metal powders for AM. For example, AP&C acquired gas atomization equipment installed at Avio Aero, both GE companies, allows production of powders from complex alloys such as titanium aluminides. Tekna Plasma Systems are increasing production to 1,000 tons annually, using a proprietary plasma atomization technology. Coogee Titanium has started producing Ti-6Al-4V with a planned capacity of 200 megatons. Equispheres Metal Powders, which has a patented atomization process claimed to give improved particle size and sphericity, has been acquired by Lockheed Martin. GKN Powder Metallurgy is working in partnership with EOS to reduce AM cost by 50% [11]. Other companies expanding powder production include PyroGenesis, Sandvik, Uddeholm, CNPC Powder Group, Metalysis, and Liberty Powder Metals (Figure 5).

To achieve a safe, airworthy, and certified component, it is critical that key process variables are controlled. Standard methods are already established for determining size, flow, and density characteristics. However, equivalent methods have not been established for the evaluation of other important variables such as sphericity, chemical composition, and thermal properties [46].

Recommendations Research to develop improved processes and investments in powder production facilities should continue to bring down the price of high-quality metal powders. There is a need for standard test methods for the evaluation of metal powder variables such as sphericity, chemical composition, and thermal properties.

Powder Reuse and Recycling

The improvement in buy-to-fly ratio is often given as a significant advantage of AM, from both an economic and an environmental perspective. Processing of aluminum alloy scrap from machining operations requires degreasing and removal of impurities using fluoride or bubbling chlorine through the melt. Recycling these materials therefore has significant financial and environmental costs. For reactive metals, such as titanium, there are currently no economical methods of recycling, although Premium AEROTEC has demonstrated

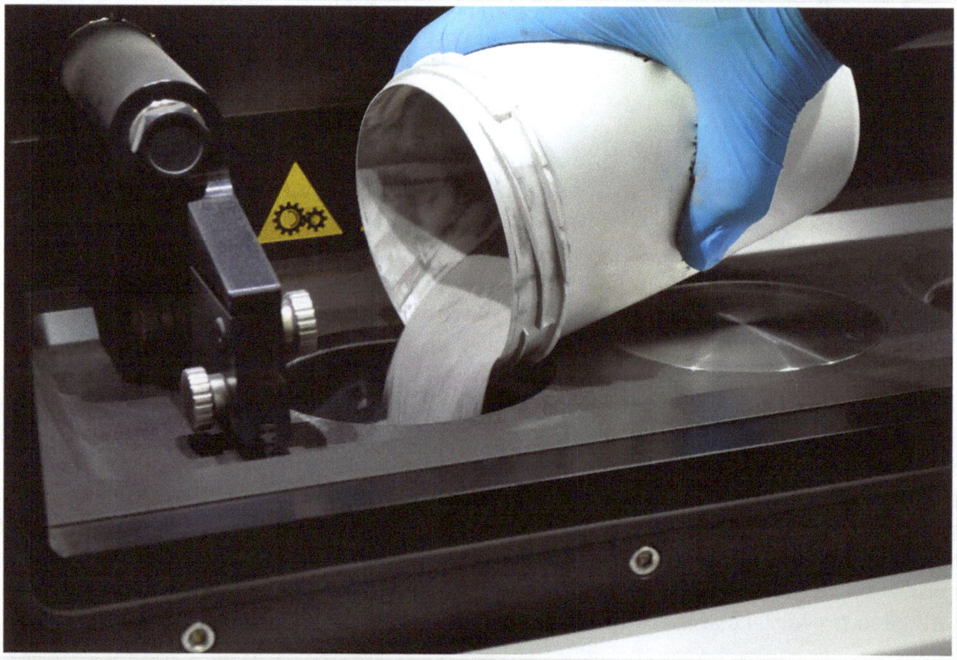

FIGURE 5. **A technician pours metal powder into the chamber of a laser sintering machine.**

MarinaGrigorivna/Shutterstock.com

that chips can be recycled into high-quality powder for AM. Furthermore, contamination of high-quality alloys may make them unsuitable for use in powder production.

It is normally assumed that unfused powder from a PBF build can be simply reused in the next build. However, most components do require some support structures which are removed as scrap and build plates, and platforms may also be considered a consumable metal product, as well as scrap produced by post-process finishing such as machining and drilling. Powders also need to be sieved, generating further scrap in the form of partially fused particles. Powder should also not be sieved and reused indefinitely due to losses related to re-melting, vaporization, oxidation, contamination, and moisture pickup during normal processing. Although, one study found that after recycling IN718, a nickel superalloy, 14 times, no measureable changes in powder quality occurred [47],

Recommendations Additional work is required to understand how many times different alloys should be reused. The need for standard tests for powder sphericity, composition, and thermal properties, as mentioned above, is also significant for determining powder reuse.

Operator and Facility Safety

Metal powders involve a significant risk of explosion, as well as inhalation risks. EB generates X-rays. EBs impinging on metals can generate X-ray hazards while Class IV lasers present eye injury and burn hazards. These laser and EB hazards are typically controlled within the confines and shielding of the chamber. Pressurized industrial and inert gases and confined spaces present their own needs for facility and operator safety.

Environment, safety, and health issues associated with handling, storage, and processing with metal powders need to be controlled. Powdered metals can be difficult to handle, store, and process. Improper storage can lead to oxidation or contamination of sources or formation of other chemical compounds that present unique hazards. Finely divided powers can be pyrophoric and when ignited burn at temperatures beyond the capabilities of handheld fire-suppression equipment. Finely divided powders can easily become airborne and contaminate surfaces creating inhalation and ingestion hazards. Enclosed processing chambers, inert processing atmospheres, sealed storage containers, specialized vacuum, filtration equipment, and proper training are but a few of the engineering and administrative controls required for the safe use of powders.

Recommendations Continued standardization of safety procedures across the industry are required.

Surface Finish

The as-deposited surfaces of PBF components feature partially fused powder. Deposition layers may be visible although surface roughness measurements are no higher when profiles are taken across the layers than when taken along a single layer. The surface texture depends on powder-particle size, melt parameters, layer thickness, and the orientation of surfaces relative to the build plate [48].

FIGURE 6. **A metal AM component rests on a printing platform covered in metallic powder medium.**

MarinaGrigorivna/Shutterstock.com

Rough-surface finish makes inspection difficult and reduces performance in terms of fatigue and fracture toughness. Although surface finish is continuing to improve the surface, machining or shot peening is often needed to improve surface finish. However, internal surfaces and some small features may be difficult to finish in this way.

Recommendations There is a need to develop new and better methods of attaining improved surface finish, which allows internal and delicate features to obtain good surface finish. For example, abrasive flow machining is being developed by Extrude Hone and electrochemical milling is being developed by Dlyte. The aim is to reach the required surface finish, on all required surfaces, as demanded by the application. Without this, topology-optimized components cannot be inspected for crack initiation defects.

Unsettled Topics Concerning DED Processes

Postprocess Machining

DED processes are inherently less accurate than PBF. Arc or plasma beams cannot be focused to less than a few millimeters, and the larger melt pools are more difficult to manipulate and control. Surface tension tends to form large rounded deposits or weld beads. Large build volumes may also use industrial articulated robots, rather than linear motion stages. These lack absolute positioning accuracy, have greatly reduced dynamic profile following accuracy, and lack the stiffness required to reject perturbing process forces.

Typically a stress relief heat treatment followed by machining and final inspection is required to meet the performance requirement of the part. Machining is required on all surfaces of structural parts as deposited surfaces can mask crack initiation sites. For a part to be fully inspected, machining is currently required. As AM quality matures and experience with AM products matures, it is likely this requirement may change.

Recommendations Data on DED part performance should be collected and the need to machine all surfaces reevaluated periodically, potentially removing significant cost from this process.

Feed Issues

Wire-fed weld pools feature the wire being consistently fed into the leading edge of the pool. Changes in direction require an articulation of the wire feed to optimize the consistency and control of the melt pool. As with all weld wire-feed applications, wire-feed irregularities due to coiling and straightening of the feed coil by the wire feeder can be an issue, as is cleanliness or other dimensional variations. NASA is also working on real-time flaw detection and FEA predictive modelling of residual stress and is working with Virginia Tech to develop software that can help design and analyze lightweight panels, such as those fabricated with EBF3.29.

Norsk's prior generation of RPD™ machines utilized articulated robotic arms to position the print head. Control of wire feed speeds and wire twist were concerns with this process, but were mitigated with the specific machine design. Overly complex machine designs (wire paths, deposition angles) are impacted by wire cast, helix, feed parameters, and speeds. It is more cost effective to optimize the machine to produce specific part-family geometries.

Recommendations Continued development of DED machines should consider geometry specific machines as well as standard industrial robots. The challenges of ensuring consistent wire-feed speed should be considered against the advantages of standardized and highly flexible machines.

Developing Hybrid Processes

Hybrid processes have the potential to overcome many of the challenges affecting the powder bed and directed energy processes. Most notably, DED is able to achieve very high deposition rates while PBF achieves greater accuracy and surface finish.

Arc and laser heat sources can be combined as a hybrid process, joining the benefits of each. Arc-assisted laser welding can combine the penetration and travel speed of a laser with the increased material deposition rates of an arc-based system. Laser beams focused within the melt region of a gas tungsten arc weld process have shown the potential to stabilize the cathode spot and arc characteristics. This combination of hybrid heat sources may further reduce the cost of future metal AM.

Recommendations Hybrid processes that combine different heat sources and feedstocks into either single machines or multistage processes should be developed and standardized

Conclusions

AM technology is maturing and in many applications is already able to produce airworthy and certified components. However, some of the most significant advantages of this technology remain underutilized due to a number of unresolved issues. For example, topology optimization for weight minimization of structural components cannot currently be widely applied because it does not consider all failure modes. By focusing on only bulk material properties, it often produces structures with many thin-walled and slender sections. These features may contain defects which could act as crack-propagation sites. However, the surface finish required for crack inspection cannot be achieved for such delicate and complex features. This means that they cannot be treated with the same confidence as thicker features. Process modelling and control are critical to AM in many respects. Process complexity means that both physics-based and empirical models cannot feasibly map all process input parameters to outputs. Testing using frozen parameter sets has enabled certified processes, but further work is required for process optimization. Improved in-process monitoring and control will also play a major role in making AM more widely used in the aerospace industry. Material properties are relatively well known for established proprietary processes, although dependence on process parameters and properties, such as fracture toughness and fatigue, could be better understood.

Costs are also currently high. Powder of sufficient quality for use in AM processes is expensive. Capital costs of PBF machines are high in relation to the rate at which these machines can deposit material. Increased machine production volumes will bring down costs, but increased deposition rates are also desirable. Step-change increases in deposition rates may be achieved through systems that maintain multiple melt pools using multiple lasers, rapidly moving EBs, or arrays of laser diodes. DED machines, on the other hand, achieve a much more economical ratio between machine cost and deposition rate, wire feedstock is also much less costly than powder. However, the requirement to fully machine all surfaces also results in high cost. Hybrid processes, which use wire feedstock to deposit the bulk of the part and powder for fine detail, could potentially enable low-cost AM production. The cost of inspecting AM components also adds significant cost, with surfaces needing to be machined for crack inspection and X-ray techniques used to detect internal defects.

Improved methods of achieving the required surface finish of fused-powder components, to a level suitable for crack inspection, will be vital to enabling topology optimization and reducing cost.

SAE EDGE™ Research Reports

SAE EDGE™ Research Reports, like the present report on "Unsettled Technology Domains in Additive Manufacturing," are intended to push further out into still unsettled areas of technology of interest to the mobility industry. SAE launches these reports before attempting to form a joint working group, let alone a CRP or a Standards Committee.

SAE EDGE™ reports are intended to be quick, concise overviews of major unsettled areas where vital new technologies are emerging. By definition, an unsettled area is characterized more by confusion and controversy than established order. Early practitioners must confront an absence of agreement. Their challenge is often not to seize the high ground but to find common ground. These scouting reports from the frontiers of investigation are intended merely to begin the process of sorting through critical issues, contributing to a better understanding of key problems, and providing helpful suggestions about possible next steps and avenues of investigation.

SAE EDGE™ Research Reports, therefore, are fundamentally distinct from the more formal working groups approach and far removed from the more mature CRP and standard's development process. They begin by convening an informal meeting of subject matter experts working in an unsettled area of technology from industry, government, and the research community. A moderated but collegial discussion ensues regarding their shared interest. This conversation attempts to identify critical issues hindering a clear understanding of important topics, preventing a common agreement on potential solutions, or slowing progress in key areas of research.

An SAE EDGE™ Research Report is then drafted and circulated for comment to meeting attendees and other interested parties. After review, the final result of this process is a report like the current examination of unsettled issues concerning AM technologies for manufacturing aerospace components.

Next Steps for Additive Manufacturing

This publication should be considered only as a first step toward clarifying the issues around AM as applied to the build of certified aerospace parts. The intention behind this and other SAE EDGE™ Research Reports is to start a dialogue among interested parties on important industry-wide topics that require further attention. The expectation is that these explorations of unsettled areas of technology will lead to the formation of working groups and, ultimately, committees that can address and resolve the issues they raise, producing a framework for developing a common vocabulary of definitions, best practices, protocols, and standards needed to support continued progress toward safer and more innovative automotive products.

The experts' meeting that gave rise to this publication demonstrated a great willingness on the part of the industry to define the terminology, procedures, and eventually the standards needed to enable AM technology to move ahead as quickly and efficiently as possible.

This SAE EDGE™ Research Report on AM identifies the following key topics for further pursuit, both through continued informal discussions among industry practitioners and through more formal working groups:

- Design for Additive Manufacturing (DfAM), development, and training in best practices are needed to establish the skills needed to fully utilize the benefits of this new technology.

- **Topology optimization** should be extended to consider uncertainties arising from variable material properties, potential for defects in thin-walled and slender sections, and to consider physics other than stress, such as fluid flow and heat transfer.
- Improved methods for **geometry creation and processing** are required that are efficient throughout the process chain, from design to certification.
- **Process modelling** and **control** are critical to AM in many respects. Process complexity means that both physics-based and empirical models cannot feasibly map all process input parameters to outputs. Testing using frozen parameter sets has enabled certified processes but further work is required for process optimization. Improved in-process monitoring will also play a major role. Continued development of mechanistic models and the application of machine learning and digital-twin technology to AM should proceed.
- **Material properties** are relatively well known for established proprietary processes although dependence on process parameters and properties such as fracture toughness and fatigue could be better understood. Open databases for materials and process-related data are needed to advance the adoption of the technology.
- **Defect detection** is currently expensive, requires surface machining, and may not detect small defects. These issues may be addressed by improved NDT or process monitoring.
- **Capital costs** of machines are high in relation to the rate at which these machines can deposit material. Increased machine production volumes will bring down costs, but increased deposition rates are also desirable.
- High **powder costs** and feedstock variation are being addressed through investments in powder production facilities and standardization of test methods. Further work is also required to better understand how powder reuse affects the process.
- Methods of improving the **surface finish** of fused powder components are required that can be applied to internal and delicate features. This is vital to enable inspection.
- The current **requirement to machine all surfaces of wire-feed components** should be reevaluated as more data becomes available, potentially removing significant costs.
- **Hybrid processes** that combine different heat sources and feedstocks into either single machines or multistage processes should be developed.

Recommendations The overall recommendations of this SAE EDGE™ Research Report can be summarized as follows:

- Methods of improving the surface finish of fused-powder components should be developed that can be applied to internal and delicate features. This will reduce cost and enable inspection of topology-optimized components. Potential methods include abrasive flow machining and electrochemical milling.
- Hybrid processes should be investigated, at the fundamental research level, which combine different heat sources and feedstocks into either single machines or multistage processes.
- Process modelling and experimental validation should continue, in particular to understand how process parameters affect microstructure development and the potential for defects in the thin-walled and slender sections seen in topology-optimized forms.
- Improved methods of designing, storing, and processing geometry are required, perhaps through the integration of parametric CAD with implicit geometry representation.

Abbreviations/Definitions

AM - Additive Manufacturing

BJ - Binder Jetting. AM process in which a liquid bonding agent is selectively deposited to join powder materials

CNC - Computer Numerically Controlled. The automated computer control of machining tools such as milling machines, grinding machines, and lathes

Crystal - A solid composed of atoms arranged in a regularly repeating 3D pattern

DED - Directed Energy Deposition. AM process in which focused thermal energy is used to fuse materials by melting as they are being deposited

Defect - An imperfection that exceeds and acceptance criteria

FAA - Federal Aviation Administration. The US body responsible for certifying that aircraft are airworthy

FEA - Finite Element Analysis - also known as the Finite Element Method (FEM). A method of simulating the physical behavior of materials that breaks larger regions down into small regions or elements whose behavior and interactions can be individually modelled relatively simple; this typically involves the solution of millions of simultaneous equations, carried out using specialized computer software

Feedstock - Material which is fused to form a part

Flaw - An acceptable imperfection

Grain structure - the arrangement of many small crystals which form a larger solid; generally smaller crystals, known as a fine grain structure, results in a stronger metal

HIP - Hot Isostatic Pressing. Process which uses heat and pressure to reduce porosity in metal parts; the part is placed in a pressure vessel filled with an inert gas such as argon and then heated

Melt pool - region of metal that has been heated above its melting temperature so that it has become a liquid; this may occur when a solid metal part is welded or when metal powder in a powder bed is melted by a high-energy source - as it cools, molten pool solidifies to form a solid part

Metallic bond - the strong bond between atoms of metallic elements that results from sharing of electrons

Microstructure - another name for the grain structure, especially the size and orientation of grains (crystals) within a metal.

MJ - Material Jetting. AM process in which droplets of build material are selectively deposited

MMPDS - Metallic Materials Properties Development and Standardization. The primary aerospace material properties database

PBF - Powder Bed Fusion. AM process in which thermal energy selectively fuses regions of a powder bed

Plasma - a state of matter in which molecules dissociate into ions and free electrons; these electrically charged particles form what is sometimes described as an "ionized gas" which conducts electricity and responds strongly to electromagnetic fields

Stair stepping - Stepped effect caused by layering in AM parts

Acknowledgments

Recognition should go first to all the participants, many of who also provided feedback on the draft version of this publication. Without their input and initiative, this SAE EDGE™ Research Report would not have been possible.

Dr. Nadia Vialas, Head of Materials and Processes, Liebherr-Aerospace

Alexander Altmann, Head of AM R&T, Liebherr-Aerospace

Claudia Jordan, EOS GmbH Electro Optical Systems

Chet Fuller, Chief Commercial Officer of Norsk Titanium

Robin Weston, AM Marketing & Product Manager, Renishaw

Jonathan Porter, AM Business Development Manager, Renishaw

Dr. Bhaskar Dutta, DM3D Technology

Dr. Joseph Flynn, Assistant Professor (Lecturer), University of Bath

John O. Milewski, Los Alamos Nat. Lab. (retired), APEX3D LLC

The authors of this document together with the SAE Team responsible for its creation join in expressing our deepest appreciation to all the individuals mentioned above.

Dr. Jody E. Muelaner
Muelaner Engineering Ltd.
SAE International masthead

Bibliographic References

1. ASTM ASTM F2792-12a, "Standard Terminology for Additive Manufacturing Technologies," 2014.
2. ISO, "Additive Manufacturing. General principles. Terminology," ISO/ASTM 529002017[cited].
3. ISO/ASTM 52911-1, "Additive Manufacturing -Design -Part 1: Laser-Based Powder Bed Fusion of Metals," 2019.
4. Milewski, J.O., *Additive Manufacturing of Metals*. Springer Series in Materials Science (Berlin, Germany: Springer, 2007), Vol. 258.
5. Adams, R., *Ion Fusion Formation: An Alternative Additive Manufacturing Approach* (North Charleston, SC, USA: SAE International, 2008).
6. Mori, D., "All-in-1 Laser Deposition Welding and Milling".
7. Airbus, "Global Market Forecast 2007 -2026," 2007, Doc.
8. Knight, H., "Additive Takes Off," *Engineer* 296(7868):36-38, 2015.
9. Additive, G., "New Manufacturing Milestone: 30,000 Additive Fuel Nozzles," 2018 [cited].
10. Stawovy, M.T., "Comparison of LCAC and PM Mo Deposited Using Sciaky EBAM," *International Journal of Refractory Metals and Hard Materials* 73:162-167, 2018, doi:10.1016/j.ijrmhm.2018.02.009.
11. Wohlers, T., Campbell, I., Diegel, O., Huff, R. et al., "Wohlers Report, in 3D Printing and Additive Manufacturing State of the Industry," 2019, Doc.
12. Administration, F.A. 8130.2H, "Airworthiness Certification of Products and Articles," 2015.
13. SAE International, "Wire Fed Plasma Arc Directed Energy Deposition Additive Manufacturing Process," AMS70052019[cited].
14. SAE International, "Titanium Alloy Preforms from Plasma Arc Directed Energy Deposition Additive Manufacturing on Substrate Ti-6Al-4V Stress Relieved," AMS70042019[cited].
15. Hassan, E., Noreland, D., Wadbro, E., and Berggren, M., "Topology Optimisation of Wideband Coaxial-to-Waveguide Transitions," *Scientific Reports* 7(45110), 2017, doi:10.1038/srep45110.
16. Alexandersen, J., Aage, N., Andreasen, C.S., and Sigmund, O., "Topology Optimisation for Natural

Convection Problems," *International Journal for Numerical Methods in Fluids* 76:699-721, 2014.

17. Dugast, F., Favennec, F., Josset, C., Fan, Y. et al., "Topology Optimization of Thermal Fluid Flows with an Adjoint Lattice Boltzmann Method," *Journal of Computational Physics* 365:376-404, 2018.
18. Liu, J., Chen, Q., Liang, X., and To, AC, "Manufacturing Cost Constrained Topology Optimization for Additive Manufacturing," *Frontiers of Mechanical Engineering*, 2019.
19. Zhang, Y., Duda, T., Scobie, J.A., Sangan, C.M. et al., "Design of an Air-Cooled Radial Turbine: Part 2 - Experimental Measurements of Heat Transfer," in *Steam Turbines*, 2018. Oslo, Norway: ASME
20. Barasuol, V., Villarreal-Magaña, O.A., Sangiah, D., Frigerio, M. et al., "Highly-Integrated Hydraulic Smart Actuators and Smart Manifolds for High-Bandwidth Force Control," *Frontiers in Robotics and AI* 5, 2018.
21. Gibson, I., Rosen, D., and Stucker, B., *Software Issues for Additive Manufacturing* (Berlin, Germany: Additive Manufacturing Technologies, 2015).
22. ASTM ISO/ASTM52915-16, "Standard Specification for Additive Manufacturing File Format (AMF) Version 1.2," 2016.
23. Consortium, M., "The Joint Development Foundation Project for creating the new file standard of 3D printing, v1.2.3," 2018.
24. Sederberg, T., Finnigan, G., Li, X., Lin, H. et al., "Watertight Trimmed NURBS," *ACM Transactions on Graphics* 27(3):79:1-79:8, 2008.
25. DebRoy, T., Wei, H.L., Zuback, J.S., Mukherjee, T. et al., "Additive Manufacturing of Metallic Components-Process, Structure and Properties," *Progress in Materials Science* 92:112-224, 2018.
26. Steen, W.M. and Mazumder, J., *Laser Material Processing* (London: Springer, 2010).
27. Society, A.W., AWS C7.1M/C7.1:2013, "Recommended Practices for Electron Beam Welding and Allied Processes," 2013.
28. Lienert, T., Siewert, T., Babu, S., and Acoff, V., *Welding Fundamentals and Processes: ASM Handbook*. Vol. 6A (West Conshohocken, PA: ASM International, 2011).
29. Lancaster, J.F., *The Physics of Welding* 2nd Edition (Oxford, UK: Pergamon, 1986).
30. Frazier, W.E., "Metal Additive Manufacturing: A Review," *Journal of Materials Engineering and Performance* 23(6):1917-1928, 2014, doi:10.1007/s11665-014-0958-z.
31. Herzog, D., Seyda, V., Wycisk, E., and Emmelmann, C., "Additive Manufacturing of Metals," *Acta Materialia* 117:371-392, 2016, doi:10.1016/j.actamat.2016.07.019.
32. Debra, T., Mukherjee, T., Milewski, J.O., Elmer, J.W. et al., "Scientific, Technological and Economic Issues in Metal Printing and Their Solutions," *Nature Materials* 18:1026-1032, 2019.
33. SAE International, AMS7003, "Laser Powder Bed Fusion Process," 2008.
34. Austin, D.C., Bevan, M.A., East, D., Brown, A.D. et al., *Microstructural Investigation and Impact Testing of Additive Manufactured TI-6AL-4V* (San Diego, CA, USA: Springer International Publishing, 2017).
35. Mohammadhosseini, A., Fraser, D., Masood, S.H., and Jahedi, M., *Microstructure and Mechanical Properties of Ti-6Al-4V Manufactured by Electron Beam Melting Process* (Leeds, UK: Maney Publishing, 2013).
36. Li, Z., Voisin, T., McKeown, J.T., Ye, J. et al., "Tensile Properties, Strain Rate Sensitivity, and Activation Volume of Additively Manufactured 316L Stainless Steels," *International Journal of Plasticity* 120:395-410, 2019, doi:10.1016/j.ijplas.2019.05.009.
37. Rice, R.C., Jackson, J.L., Bakuckas, J., and Thompson, S., *The Metallic Material Properties Development and Standardization (MMPDS) Handbook* (Department of Transportation Federal Aviation Administration, 2003), Doc.
38. US Department of Defense, Metallic Materials and Elements for Aerospace Vehicle Structures, *Military Handbook* (US Department of Defense, 1998), Doc Ref *MIL-HDBK-5H*
39. Lewandowski, J.J. and Seifi, M., "Metal Additive Manufacturing: A Review of Mechanical Properties," *Annual Review of Materials Research* 46:151-186, 2016, doi:10.1146/annurev-matsci-070115-032024.
40. Murr, L.E., Martinez, E., Amato, K.N., Gaytan, S.M. et al., "Fabrication of Metal and Alloy Components by Additive Manufacturing: Examples of 3D Materials Science," *Journal of Materials Research and Technology* 1(1):42-54, 2012, doi:https://doi.org/10.1016/S2238-7854(12)70009-1.
41. Gong, H., *Generation and Detection of Defects in Metallic Parts Fabricated by Selective Laser Melting and Electron Beam Melting and Their Effects on Mechanical Properties* (Louisville: Department of Industrial Engineering, University of Louisville, 2013).
42. Everton, S., Dickens, P., Tuck, C., and Dutton, B., "Using Laser Ultrasound to Detect Subsurface Defects in Metal Laser Powder Bed Fusion Components," *JOM* 70(3):378-383, 2018, doi:10.1007/s11837-017-2661-7.
43. Davies, S., "EOS to Introduce LaserProFusion Polymer 3D Printing Technology at Formnext," *TCT Magazine*, 2018.
44. Fenn, A.J., Cooley, G., Fray, D., and Smith, L., "Exploiting the FFC Cambridge Process," *Advanced Materials and Processes* 162(2):51-53, 2004.

45. Marshall, L.B., Benson, L.L., Weston, N.S., Repper, S.E. et al., *Development of Titanium Alloy Powders Produced via the Metalysis Process* (Milan, Italy: European Powder Metallurgy Association (EPMA), 2017).
46. Cooke, A. and Slotwinski, J., "Properties of Metal Powders for Additive Manufacturing: A Review of the State of the Art of Metal Powder Property Testing," 2012, NIST, Doc Ref NISTIR 7873.
47. Ardila, L.C., Garciandia, F., González-Díaz, J.B., Álvarez, P. et al., "Effect of IN718 Recycled Powder Reuse on Properties of Parts Manufactured by Means of Selective Laser Melting," *Physics Procedia* 56:99-107, 2014, doi:https://doi.org/10.1016/j.phpro.2014.08.152.
48. Triantaphyllou, A., Giusca, C.L., Macaulay, G.D., Roerig, F. et al., "Surface Texture Measurement for Additive Manufacturing," *Surface Topography: Metrology and Properties* 3(2), 2015, doi:10.1088/2051-672x/3/2/024002.

Contact Information

EDGEresearch@sae.org.

The authors of this document together with the SAE Team responsible for its creation join in expressing our deepest appreciation to all of the individuals who contributed.

www.ingramcontent.com/pod-product-compliance
Lightning Source LLC
LaVergne TN
LVHW070941210926
848482LV00003B/274